Everything You Need To Know About Your Pellet Stove –

and more*!*

---

written by

Andrew Mercaldo

Pendale House Press
New York

Everything You Need To Know About Your Pellet Stove and More

First Edition

Character Illustration by Justin Contois

Published by: Pendale House Press – New York
pendalehouse@yahoo.com
PO Box 995 Staten Island, New York 10306

All Pendale House Press books are printed in the United States of America

Table of Contents

ABOUT THE AUTHOR

Andrew Mercaldo describes himself as part of the last generation of "machine men."

Although he has an academic background (his career includes thirty years as a title attorney for a New York title insurance company, and fifteen years as the pastor of a small Vermont church), there was always the love of anything mechanical.

About the same time of his retirement from ministry, he purchased a pellet stove and became fascinated with its simple mechanical function. This prompted him to pursue certification as a pellet stove specialist, and start a business servicing pellet stoves.

This book grew out of awareness that most pellet stove owners knew very little about their stoves and the care they required.

The author lives in Vermont with his wife, Betty. The couple was married in 1960.

The author is a National Fireplace Institute Certified Pellet Stove Specialist (#167044)

A WORD FROM THE PUBLISHER

When Andrew Mercaldo approached us with the idea of publishing a book about "Pellet Stoves," we were indeed intrigued; but had to answer a few questions before we could take on his project.

Don't pellet stoves come with easy to understand instructions for installation, operation and simple repairs?

Is it really possible for a homeowner with little mechanical knowledge to install, trouble shoot and service a pellet stove?

These questions compelled us to read his manuscript to see if pellet stove owners could use such a manual.

A simple survey indicated there was a real need for a simple and practical guide to purchasing, installing, maintaining and servicing a pellet stove unit.

We also discovered something else! Pellet stoves are fast becoming a "heating sensation" across the United States with stoves being purchased as an alternative to oil, gas, coal and electric heating units.

So here is *EVERYTHING YOU NEED TO KNOW ABOUT YOUR PELLET STOVE...AND MORE!*

Please enjoy our large print edition of this book.

Is there a pellet stove
in your future?
After reading this book, you
may be joining the millions who
are enthusiastically embracing
this eco-friendly option for home
heating.

PREFACE

PELLET STOVES are becoming more popular and we are hearing positive reports from stove owners.

In many situations, installing a pellet stove is an effective way to control heating costs and enjoy a warm home during the cold winter months. Yet, these stoves are not suitable for every home or the right choice for every homeowner.

My purpose in writing this booklet is to inform those who are considering the option of installing a pellet stove so that they can make an informed decision as to whether it is right for them.

I am grateful for the training I received from the National Fireplace Institute, and hope that this book does justice to that training.

Andrew Mercaldo, Author

INTRODUCTION

Some Basic Considerations

Let's face it! You would not be thinking of buying a pellet stove if you were not concerned about the cost of heating your home. The intent of this booklet is to help you make the decision whether a pellet stove is right for your home. I wish this information were available when I bought my first pellet stove.

Twelve years ago I had our house built and opted for a forced - air furnace unit to heat it. At the time, a friend advised me to include a masonry chimney in the construction. Propane was selling for about a dollar a gallon and I saw little need to plan for any alternative heating. It did not take long for the price of propane to steadily increase. The cost became a real concern and rumors of shortages pushed me over the edge. We did not have a chimney – propane furnaces require only venting. Pellet stoves also do not require a chimney and it seemed to be the way to go.

Without any real research on my part, I ordered a pellet stove from an online vendor. I did not intend to replace the propane unit; it would remain the back up. Since I'm a "do-it-yourselfer," I thought it would be an easy task.

When the stove arrived, I enthusiastically set out to install it with the prescribed venting.

From the start, it proved to be a difficult and rather complicated job. I knew little about these appliances and had not done a good job in planning for the installation. I met with one problem after another. Finally, after many starts, I got the pellet stove to operate according to the manufacturers specifications. When I felt the heat from the stove for the first time, I went back to the old adage…"If only I knew then what I knew now!" Anyway, it was up and running.

To continue, all new pellet stoves do come with an owner's manual. These instruction booklets are very helpful and anyone with basic construction skills, who carefully follows the instructions within the manual, can do an acceptable job of installation. The problem is that you get the manual only when the stove arrives.

<u>Planning for the installation of a new pellet stove should take place long before you actually make the purchase.</u>

There are installation and maintenance issues in purchasing and owning a pellet stove, so I will try to be as clear as possible in describing these in this booklet. Above all, I do not want to discourage you from buying a pellet stove for your home.

My wife and I love our pellet stove. Unlike our original forced air propane furnace, the pellet stove sends out a steady soft heat; and while there are maintenance issues, the benefits as they relate to cost and comfort, far outweigh the extra effort of maintaining your pellet stove. I am yet to meet a pellet-stove owner who would give up his or her stove.

The content of the following chapters will follow a logical - chronological order, and answer the questions you should ask before you make the final decision about pellet-stove ownership.

FIVE KEY QUESTIONS

1. Is a pellet stove feasible for your home?
   (Can it be installed and comply with the codes, standard and manufacturer's instructions?)

2. What exactly is a pellet stove?
   (How does it work?)

3. What are pellets?
   (Are all pellets the same?)

4. What kind of care and maintenance does a pellet stove require of the owner?

5. Where can you buy your pellet stove?
   (Should your purchase be made at a pellet stove store that specializes in selling, installing and servicing pellet stoves, or should you purchase it at a place like a "big box hardware store" or on-line, and do the work yourself?)

The following chapters will deal with these five questions

Chapter One

Is a Pellet Stove feasible for your home?
(Can it be installed and comply with the codes, standards, and manufacturer's requirements?)

A fundamental point must be made before we talk about "feasibility."

If you are thinking of replacing a wood-burning stove with a pellet stove, you should be aware of the differences in heat production of these two appliances. Depending on the amount of fuel, a wood-burning stove will burn much hotter than a pellet stove. If you now have a wood-burning stove and use all the heat the stove produces, you will be disappointed with a pellet stove.

In addition, it is recommended that you keep your present heating unit operating. While you may want to use your pellet stove as your primary source of heat, it is a good idea to keep your present central heating unit as a backup. If you plan to be away from your home for a few days during the winter, that old heating system will come in handy.

As you begin planning, it is essential to know the LOCAL CODES and BUILDING PERMIT REQUIREMENTS of your local municipality, and any state requirements. Every town and state has different codes.

Your local town building department official can help you understand the local requirements.

Explain what you plan to do and ask for their advice. A building permit may be required. Go online and locate your state's web page. Most are interactive and you will be able to ask if there are any special state requirements for installation of pellet stoves.

If necessary, call your state official and ask for help. If new electrical outlets are required for the proposed stove location, the work may have to be done by a licensed electrician. Be focused, persistent and be sure you understand the local and state requirements before you go any further.

The codes and standards affecting wood burning appliances (pellet stoves are considered in this class) are set out in the International Residential Code (IRC) and National Fire Protection Association (NFPA).

The National Fire Protection Association
(NFPA) is a trade association that sets codes
and standards that are aimed at fire protection.
Many parts of the country now recognize
the International Residential Code (IRC).
The NFPA and IRC hold to similar standards.

The owner's manual that comes with pellet stove, will tell you all that you need to know about these codes, standards and requirements of the particular stove manufacturer.

Again, you must determine if a pellet stove can be installed in your home in compliance with the codes, standards and manufacturer's requirement before you make your purchase.

You may have noticed that pellet stoves can be vented through an outside wall and below the roofline. Pellet stoves do not require a conventional chimney and can be vented straight out from the stove through the outside wall. Some go straight out and terminate. Others go straight out and continue with a vertical section of venting that runs up along the outside wall. Some have the vertical section along the inside wall and go through the outside wall above the pellet stove. This feature of pellet stoves makes them very attractive when a building does not have a chimney or where one cannot be easily constructed.

(Note the following requirements as you do your pre-planning. The following are shown for informational purposes only. You may check the text of NFPA and IRC online for the current codes).

The TERMINUS of the vent outside of the building must be:

a. Not less than 3 feet above any forced air inlet, located within 10 feet.

b. Not less than 4 feet horizontal from doors, or operating window (below or beside); or 1 foot above any window or gravity air inlet into the building.

c. Not less than 2 feet from any adjacent building; and not less than 7 feet above grade when located adjacent to public walkways.

d. Not less than 2 feet below eves or roof overhangs.

e. Not less than 1 foot from any wall of combustible material.

2. The terminus of the venting must be located so that flue gases will not jeopardize the health of any person, overheat combustible material or enter any building structure.

3. Mechanical forced venting systems and any vent system under positive pressure when in operation (a characteristic of pellet stoves) shall be <u>gastight</u> to prevent any combustion gases from entering the building.

To illustrate the importance of taking the codes and standards into consideration I relate the following experience.

I was called to install a new pellet stove that was purchased at one of the big box stores. When I arrived, the owner brought me into the living room where he wanted the stove to be installed. All of the outside walls were covered with windows. There was not one spot where I could exit the venting and be four feet from an operating window. There was one space of wall about one foot wide between two windows, and I suggested that the owner permanently seal those windows, making them inoperable. Anticipating cutting and venting through the wall to the outside, I tapped the area with my knuckles and the solid sound of a beam was heard.

As I studied the room, it was obvious that there was a supporting beam in that space. I told the owner I would not get involved with the installation and declined the job. In his case, it would take some serious reconstruction work to accommodate the pellet stove in that room.

NOTE: The owner's manual, which came with the new stove, did set out the "code requirements" and manufacturer's specifications for installation, but it was too late for the owner to plan properly.

MOBILE HOME INSTALLATION

If you are planning to install the pellet stove in a mobile home, be advised that the Department of Housing and Urban Development (HUD) has special requirements for installation of wood burning stoves (pellet stoves come under this category). I suggest you go on line and search – "HUD REQUIREMENTS FOR THE INSTALLATION OF WOOD-BURNING STOVES IN

A MOBILE HOME," and you will find those specifications.

HUD requirements will also be found in the manufacturer's owner's manual; but again, you should know those requirements before you make a purchase.

Manufacturers generally do not recommend that pellet stoves be installed in bedrooms.

Another factor to take into consideration is the requirement that special pellet stove venting, known as PL or L venting be used in the installation of your stove. These vent pipes, tested to UL standards, are double walled with a stainless steel inner wall. Some brands have interlocking sections that provide a gas seal. With other brands, the sections must be sealed with a high temperature sealant. PL venting provides an insulating value to reduce clearance to combustibles.

Depending on the manufacturer or the country (US or Canada), clearance to combustibles will vary from 1 to 3 inches from venting. PL venting is significantly more expensive than single wall venting. Single wall venting violates NFPA-211 and should not be used with pellet-stove installations, unless approved by the manufacturer for special installations.

(I will talk more about venting in a later chapter)

The pellet stove must be located to provide adequate clearance between the body of the stove and any combustible materials. The owner's manual will give the distances that must be maintained between the stove and walls or any other combustibles.

There will usually be a requirement for a floor covering under the stove made of non-combustible material. The relationship between the size of the floor covering and the dimensions of the stove will be listed in the owner's manual.

Once you have decided on the make and model of your stove, go online and review the manufacturer's owner's manual.

Most manufacturers make their manuals available online. Determine if the stove can be installed with the codes and standards, and all of the manufacturer's special requirements.

ONE MORE POINT FOR CONSIDERATION

YOU MUST NOTIFY YOUR HOMEOWNER'S INSURANCE COMPANY OF YOUR PLAN TO INSTALL A PELLET STOVE.

<u>Insurance companies consider the installation of a pellet or wood stove to be an additional risk.</u>

Don't install - then call. Call – then install!

There may be a premium adjustment for your policy. With all the natural disasters we have been experiencing as a nation, insurance companies are paying out huge claims and are strapped for cash. Don't give them any opportunity to deny a claim based on an undisclosed risk.

If you plan to install the stove yourself, be prepared to prove that it was properly installed.

One way to satisfy your insurance company is to go online to the NATIONAL FIREPLACE INSTITUTE web page and search the site for a NFI CERTIFIED PELLET STOVE SPECIALIST in your area who would be willing to inspect your installation and certify that it is compliant. I assume there will be a fee for this service.

Chapter Two

What is a Pellet Stove?

A pellet stove is a solid-fuel burning heat producing appliance, incorporating a mechanical fuel-feed system that delivers small measured batches of pellets to a combustion chamber.

Typically, combustion gases and smoke produced in operations are exited mechanically by means of an electric fan directed into venting. There are a few manufacturers that offer "natural draft" pellet stoves, but more typically, pellet stoves will employ an electric combustion or draft fan.

The best way to understand what a pellet stove is, and how it operates, is to consider each of the following operating components separately:

A. Fuel Delivery

B. Combustion Air Delivery

C. Heat Transfer

D. Ash Collection

E. Venting

Before considering each component, it would be well to mention the CONTROL BOARD and safety features. The various operating components are synchronized by a Control Board that is "the brains of the operation."

Typically, the Control Board has several automatic features and some that are controlled manually. Control Boards differ according to the specifications of individual manufacturers. Your Owner's Manual will guide you in the basic operation of the stove.

Pellet stoves, if installed and maintained according to the manufacturer's instructions in the owner's manual, are relatively safe. The fire and flames in the pellet stove are confined in a sealed chamber. (Needless to say, depending on the design of the individual stove, the surface may get very hot, and will burn the skin if metal surfaces are touched. Children should be kept away from operating pellet stoves).

Flames will be visible through the glass on the combustion chamber door but will not leap from the stove as they can from a conventional fireplace.

There are also switches that will cut off the fuel supply if the pellet stove is running too hot or not running safely. I emphasize, that when installed and maintained according to the manufacturer's instructions in the owner's manual, a pellet stove is relatively safe when compared to open wood burning fireplaces. If these instructions are ignored or compromised, the pellet stove will be unsafe to operate.

As we discuss the operation of a pellet stove, keep in mind that the three ingredients needed for combustion are fuel, air, and ignition.

Now, let us look at the individual components of a pellet stove:

Operating Components

a) The Fuel Delivery Component

The fuel delivery component is comprised of a HOPPER for storage of the pellets, the AUGER TUBE AND AUGER for the delivery of the pellets, and the BURN POT in the combustion chamber where the burning takes place.

The size of the hopper differs depending on the stove make and model. Typically, the storage area of the hopper ranges from 40 to 120 pounds of pellets. Pellets are usually sold in 40 lb bags or in bulk. Since 2004, pellet stove manufactures must include a lid switch for the lid of the hopper compartment. The switch cuts off current to the auger motor when the lid is in an open position. This is a safety feature to prevent fingers from being caught in the auger of an operating stove.

The hopper is connected to the auger tube. Pellets are typically gravity fed into the tube. The auger itself looks like a "corkscrew."

The auger is powered by a low rpm electric motor that rotates the auger, delivering pellets to the burn pot. The auger rotates intermittently at a specific rate according to the heat range setting on the control board.

The auger turns with off-on cycles. The auger may be powered *on* for 2 seconds and *off* for 10 seconds. The longer the on cycle, the greater the number of pellets that are fed to the burn pot and the higher the temperature.

There are two basic operating designs for augers. One is the "top feeder," and the other is called a "horizontal feeder."

In the top feeder, the auger and tube are set on an inclined plane and pellets are lifted by the corkscrew action of the auger and dropped into the burn pot. In the horizontal design, the auger and auger tube are set in a horizontal position and the pellets are pushed by the auger action into the burn pot.

When choosing a pellet stove, you should ask about the auger design – whether a top feeder or horizontal feeder. Take the time to do some online research and learn about the pros and cons of each.

The burn pot is located in the combustion chamber and is considered part of the fuel delivery component. Generally speaking, the burn pot is where ignition and combustion of the pellets takes place. Burn pots are usually made of hardened steel and can be square, rectangular or round, and are 4 to 6 inches wall to wall. Burn pots or "boxes" are characterized by air holes that allow air to enter and mix with pellet fuel aiding the combustion.

FIGURE 1

FUEL DELIVERY COMPONENT

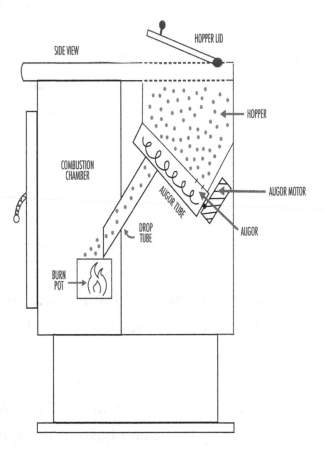

b) The Combustion Air-Delivery Component

Almost all manufacturers of pellet stoves that are sold today in the US and Canada use an electric combustion or draft fan to mechanically draw air into the combustion chamber and burn pot. The same fan, by the same action, blows the combustion gases, ash soot and smoke into the venting system.

Natural draft pellet stoves are popular in Europe and there are a few manufacturers in the US producing this type of pellet stove. If you have an interest in a natural draft pellet stove, you can get information online.

My discussion in this chapter will focus on the mechanical electric combustion or draft fan.

As we consider the combustion air delivery component, we will follow the flow of air from the air intake and ending with the combustion or draft fan. The movement of air from the air intake to the exhaust of combustion gases creates a negative pressure or atmosphere in the pellet stove.

The air intake is usually located at the back of the stove and depending on the instructions of the manufacturer, may use room air or may require an air inlet tube to run from the stove, through a wall, to outside air.

By the action of the combustion or draft fan, air will be drawn into the air inlet and will run through a channel or tube to the burn pot where air will aid in combustion. Unburned particles, combustion gases, soot and some ash, are then drawn into the combustion fan and blown out into the venting.

Depending on the make and model of the pellet stove, the amount of air allowed to enter the air inlet is either automatically controlled by the Control Board or is manually set with an air damper. Again, depending on the make and model, the speed of the combustion fan motor is either controlled by the control board or manually set by speed-control buttons or dials.

The amount of air intake is coordinated with the speed of the auger feed motor, so that the amount of air drawn into the burn pot is sufficient for the amount of pellets to produce an efficient flame and burn.

The negative pressure or atmosphere in the stove helps insure that air will be drawn into the stove, not only from the air inlet, but from any other air openings, and lessens the possibility of combustion gas leaking into the living space. (Outside air is always being drawn into the combustion chamber by the action of the draft fan).

The negative pressure or atmosphere must be maintained for safe efficient operation of the stove. There are vacuum safety switches which detect any fall in negative pressure and cut off electric current to the auger feed motor, stopping fuel from entering the burn pot leading to the shut down of the stove.

FIGURE 2

COMBUSTION AIR DELIVERY COMPONENT

OUTSIDE AIR IS DRAWN INTO THE BURN POT WHERE IT SUPPORTS COMBUSTION. COMBUSTION GASES ARE BLOWN INTO VEN

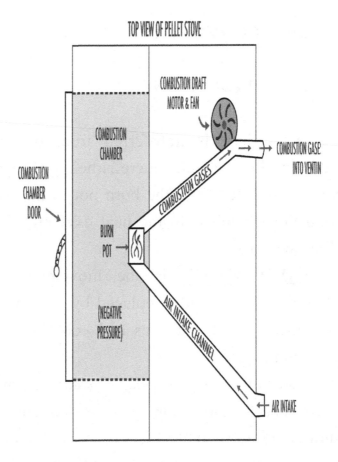

TOP VIEW OF PELLET STOVE

c) The Heat Transfer Component

We operate a pellet stove to heat our living space. Generally, heat is transferred by three means:

1)   **CONDUCTION**

2)   **CONVECTION**

3)   **RADIATION**

When heat is transferred from one surface to another by direct contact, there is heat **CONDUCTION**. Heat is conducted from the burn pot to the walls of the combustion chamber, to the heat exchanges and to the wall of the stove.

CONVECTION is the movement of lighter, heated air upward that is replaced by heavier cooler air, which then heated and rises and continues the natural convection cycle.

Mechanical or *forced convection* employs an electric fan to move air over heated surfaces. Both natural and forced convection operate in the heat transfer component of pellet stoves. Heated air inside the stove body rises and is replaced by cooler air which in turn is heated and rises.

Pellet stoves employ room or convection fans to blow cooler air through, over or around heat exchanges heating the air and blowing the heated air into the living space. Heat exchanges can be hollow tubes, fins or air chambers that are heated by the pellet fire.

Subject to the make and model of the stove, *the speed of the convection* or room fan is either controlled by the control board, or manually set by a button or dial. The convection or room fan, not only delivers warm air into the living space, but also cools the stove, preventing overheating.

As mentioned before, pellet stoves have safety switches or *sensors* that will cut off electric current to the auger-feed motor if the stove gets too hot. Without fuel in the burn pot the stove will shut down.

RADIANT heat is the result of infrared energy converted to heat energy on contact with a solid object – including people. In an operating pellet stove, radiant infrared energy travels through the air from the glass in the combustion chamber door, and from other heated surfaces of the stove to solid surfaces and objects – including people.

The air between the source of the radiant heat and solid surface is not noticeably heated. A good example of this is the sun on a cold day. Standing outdoors, it is the sun that warms us; yet if we were to check the temperature of the air in the sunlight, and the air in the shade, it would be about the same. Radiant heat, from the sun, heats solid surfaces and objects and people. The same phenomenon is at work in the operating pellet stove.

Virtually all brands of pellet stoves allow for the installation of wall thermostats, thus by-passing the manual heat controls on the control board.

The owner's manual gives careful instructions on the installation of external thermostats.

Illustrated on the following page:

FIGURE 3

HEAT TRANSFER COMPONENT

ROOM AIR IS BLOWN THRU HEAT EXCHANGE TUBES

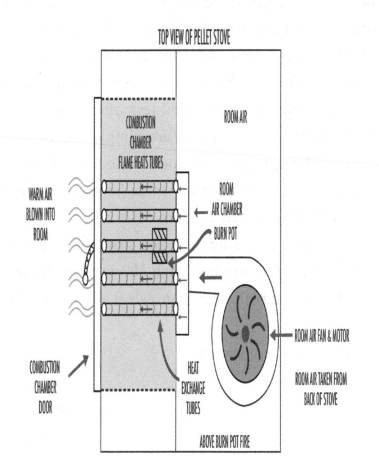

TOP VIEW OF PELLET STOVE

COMBUSTION CHAMBER
FLAME HEATS TUBES

ROOM AIR

WARM AIR BLOWN INTO ROOM

ROOM AIR CHAMBER

BURN POT

ROOM AIR FAN & MOTOR

COMBUSTION CHAMBER DOOR

HEAT EXCHANGE TUBES

ROOM AIR TAKEN FROM BACK OF STOVE

ABOVE BURN POT FIRE

d) The Ash Collection Compartment

Combustion in the burn pot produces wood ash.

Ash will be heaviest in the burn pot area of the combustion chamber. Generally, there are two ways ash is collected in pellet stoves. Some stoves have ash boxes or drawers located below the combustion chamber. In others, there are ash pans in recessed areas in the bottom of the combustion chamber below the burn pot. Ash boxes or drawers hold considerably more ash than ash pans and have to be emptied less often than the ash pans. The ash boxes or drawers are simply removed from the stove and emptied out of doors. The ashes in the ash pans in the combustion chamber must be scooped out by hand or emptied by an ash vacuum.

When choosing a pellet stove, ask about the ash collection compartment. Ash drawers or boxes require much less maintenance than ash pans in the combustion chamber.

Illustrated on next page

FIGURE 4

ASH COLLECTION COMPONENT

WITH ASH **DRAWER**

SIDE VIEW

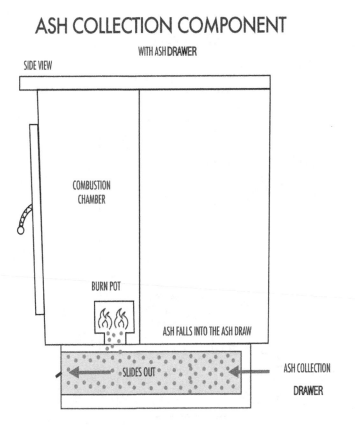

FIGURE 5

ASH COLLECTION COMPONENT WITH ASH PAN

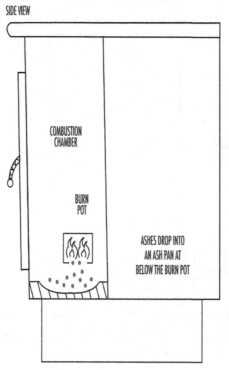

SIDE VIEW

COMBUSTION
CHAMBER

BURN
POT

ASHES DROP INTO
AN ASH PAN AT
BELOW THE BURN POT

ASHES MUST BE REMOVED FROM ASH PAN AT FREQUENT INTERVALS

e) The Venting Component

Venting must be understood as an integral part of the operation of a pellet stove. We could say that the stove begins at the air intake and ends at the terminus of the venting where the combustion gases are released into the open air.

The National Fire Protection Association requires PL or L type pellet stove venting for all pellet stoves. PL or L venting comes in various lengths and are double walled with a stainless steel inner wall. If the sections of venting are not self-sealing, they must be sealed at every connection with high temperature sealant. The double walled PL or L venting provides an insulating factor so that the venting can be closer to combustible material.

As stated earlier, the clearance between PL and L venting and combustibles must be at least 1" (US) or 3" (Can). I always opt for the 3" clearance when installing venting.

It is important to note that the total length of horizontal and vertical venting from the stove exhaust to the outside vent terminus will be specified by the manufacturer. The number of elbows and clean-out Ts will also be specified. These specs must be complied with if the stove is to operate efficiently.

After deciding on the particular make and model of pellet stove, and before you make your purchase, go online and study the owner's manual of the particular stove and study the venting requirements. The venting requirements of some stoves may be too stringent for your particular installation.

If you plan to use an existing manufactured chimney, it is essential that you call the technical department of the manufacturer of the stove you intend to buy or the retailer where you will make your purchase and describe in detail the make and model of the manufactured chimney and what you intend to do. Use of existing manufactured chimneys for pellet stoves is generally not permitted. Check it out before you make your purchase and begin installation.

If you plan to exit your venting into an existing masonry chimney, be aware that no other appliance or stove can use the same chimney. If you intend to install the pellet stove yourself, it would be a good idea for safety sake, to call in a <u>certified chimney sweep</u> to come and examine the chimney to see if it is in good condition and is suitable for the pellet stove use.

Here again, it would be advisable to call the technical department of the stove manufacturer or retailer and describe to them exactly what you are planning and get their approval.

Some manufacturers require that a liner be installed in masonry chimneys before installing your stove, while other manufacturers do not have such a requirement

It would be a good idea to investigate before you make your purchase. Venting requirements vary from manufacturer to manufacturer, and may determine which stove you should buy.

In a later chapter, I will talk about the option of buying from a local pellet stove retailer or from a big box store or online store. There are brands of stoves that are sold only through local pellet stove retailers, and other brands that are sold directly to the public through big box stores or online stores.

Those manufacturers that sell only through local retailers will direct any inquiry about installation to the retailer. Those manufacturers that sell directly to the public have technical departments to handle any inquiry about installation.

(See next illustration of venting component)

Figure six

VENTING COMPONENT
SHOWING INTERIOR VENTING WITH EXIT OVER STONE

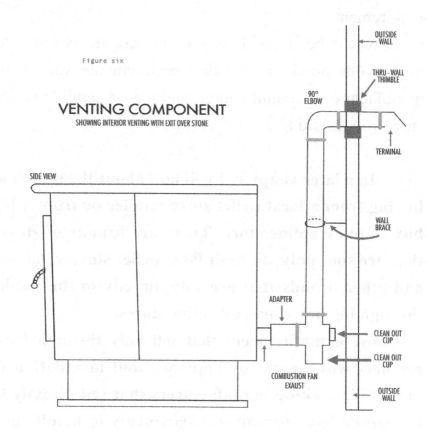

SIDE VIEW

OUTSIDE WALL

THRU - WALL THIMBLE

90° ELBOW

TERMINAL

WALL BRACE

ADAPTER

CLEAN OUT CUP

CLEAN OUT CUP

COMBUSTION FAN EXAUST

OUTSIDE WALL

Chapter Three
What are Wood Pellets?

Without going into the details of the manufacture of wood pellets, it is enough to know that the pellets are processed from waste wood and sawdust from commercial operations. The wood is machine-pressed into pellets approximately one quarter of an inch in diameter and one half to three-quarters of an inch in length.

The Pellet Fuel Institute (PFI) is a voluntary trade association that sets standards for wood pellets. The finished pellets are usually packaged in 40 lb. plastic bags and should display a stamp and seal of the PFI – indicating the grade of the pellets. Grades range from "standard" to "premium." It is advisable to always choose premium grade pellets as certified by the PFI.

I have found that different pellets of different manufacturers burn differently even though they are graded "premium." Try different brands of pellets and find the one that burns best in your pellet stove.

ALTERNATIVE FUELS

I should mention here that some pellet stoves are designed to burn non-wood pellet fuels such as dried corn kernels or cherry pits, while others are designed to burn both wood pellets and dried corn kernels. If you are thinking about going the non-wood pellet or mixed route, investigate the availability and care of these alternate fuels. Remember corn is food and it will draw insects and rodents.

STORAGE

Wood pellets must be stored in a dry place. Once pellets get wet or even damp, they fall apart and are useless. As you plan, decide where you will store your pellets. Some owners opt to pick up pellets daily or weekly. That will work most of the time, but in cold winters, wood pellets become very scarce in January, February and March.

I usually buy three tons of pellets in the summer so that I can be sure of a supply.

A question frequently asked about pellets is their cost relationship to other fuels. The cost ratio of wood pellets to propane, fuel oil and electric is very favorable for pellets. You can search online and find sites where the cost comparisons are made between the wood pellets, propane, fuel oil, electric and natural gas. These ratios change as prices fluctuate.

I found one site that compared a 40 lb. bag of pellets to propane, oil and natural gas. It held that a 40 lb. bag of pellets has the same heat content as 2.3 gallons of fuel oil; 3.5 gallons of propane; 3.2 therms of natural gas; and

94 kWh of electricity. Prices of these fuels are always fluctuating, but if you do the math, you will find that pellet stoves are cost effective.

Chapter Four

Care and Maintenance of
Your Pellet Stove

Wood pellet stoves are not clean burning appliances. They produce a lot of ash and soot in the stove, and in the venting. They require maintenance if they are to operate safely and efficiently. Here again, the owner's manual will set out a daily, monthly and yearly cleaning schedule. It is a good idea to have a high quality dust mask, and an ash vacuum on hand when you service your stove.

Daily maintenance

The daily cleaning focuses on the glass in the combustion chamber door; the burn pot; heat exchanges and refilling the pellet hopper. This process takes about 5 - 10 minutes.

Monthly maintenance

The monthly cleaning goes a little deeper into the stove's interior. The owner's manual will give instructions on the simple disassembly and service required. The monthly maintenance will take a little longer than daily servicing.

Annual maintenance

At the end of the heating season, as you shut down the stove for the warm months, a more intensive service maintenance is required. Here again, the owner's manual will give you specific instructions for the end of season maintenance.

Typically, for the annual service and maintenance, the stove owner will be instructed to do the following:

1. **Empty the pellet hopper.**
2. **Let the stove run and burn out all of the pellets in the auger tube, and allow the stove to shut down by itself.**
3. **Unplug the stove for the season.**
4. **Vacuum the hopper of all pellet dust.**
5. **Make sure the air intake is clear.**
6. **Vacuum the combustion chamber- including the heat exchanges. There are usually baffles that must be removed and ash traps that are exposed and must be vacuumed.**
7. **The burn pot must be cleaned, air holes cleared and creosote build-up scraped**

clean. (This is also an important part of the daily and monthly maintenance).

8. The exhaust channel or tube running from the combustion chamber to the combustion or draft fan must be cleared of all ash buildup and any creosote scraped clean.

9. The combustion or draft fan housing must be opened and the motor fan blades scraped, removing any soot or creosote. (A combustion fan with crusted fan blades will slow the airflow and rob efficiency).

10. Vacuum the fan housing and the exhaust passage to the venting.

11. Empty the cleanout Ts and using a pellet stove vent brush, brush out all parts of the venting.

This yearly maintenance and cleaning usually takes about 2 hours to complete.

Pellet stove manufacturers usually recommend that a Certified Pellet Stove Specialist do the annual maintenance, but most stove owners can to this servicing by carefully following the instructions in the owner's manual. If you decide that you need the help of a specialist to do this work or for any other repairs, you can find one in your area by visiting the webpage of the National Fire Place Institute (www.nficertified.org)

Different pellet stoves require different levels of attention and maintenance. Different brands of pellets also affect how often your stove will need cleaning and maintenance.

As you work with your stove, you will find your own routine for keeping it in good working order.

It has been my experience, working in the field servicing and repairing pellet stoves, that the chief cause of problems is failure to properly install and maintain the stove according to the instructions in the owner's manual.

Cleaning and maintaining your pellet stove may seem like a lot of work, but the reward of a dependable and efficient stove burning brightly and sending out warmth on a cold winter day is worth all the effort.

Chapter Five

Where Should I Buy My Pellet Stove?

There are basically two approaches to the purchase of a new pellet stove: You can go to a pellet-stove retailer where stove specialists will sell, install and maintain the stove you purchase; or you can purchase a stove from a big box store or online store and do the installation, and maintenance yourself. Let us look at the benefits and drawback of each.

If you go to a pellet stove store and deal with professionals, you can be sure that your stove will be installed in compliance with all codes, standards and manufacturer's requirements. They will also deal with the local Building Department officials to get all the proper permits etc. You can further be assured that your homeowner's insurance company will not question the installation.

Pellet stove retailers offer maintenance plans to purchasers, and you will have the assurance that if the stove needs repair or maintenance, the servicing will be done by a specialist.

Upon installation, you will have a dependable and efficient operating pellet stove. The only drawback is the cost.

The cost of even a modest pellet stove with installation will run between $5,000 and $6,000 or more depending on the complexity of the installation. If you can afford the cost, the best way to go is to deal with professionals.

If you are a "do-it-yourselfer," and decide to make your purchase at a big box store or online, a modest pellet stove, with the necessary venting, will run about $2,000 to $3,000. Do your homework before your purchase and make sure that your installation will comply with the requirements of the local officials and all codes, standards and the manufacturer's requirement.

Even if you plan to do the installation yourself, it would be a good idea to get the help of a NFI Certified Pellet Stove Specialist to help you plan the installation. Again, you can get the name of a local NFI Certified Pellet Stove Specialist by going to the web page of the National Fireplace Institute (NFI) at the following site:

www.nficertified.org

Your homeowner's insurance company will question whether the stove and venting has been installed in compliance with the codes etc.. A letter from the NFI

Certified Specialist, stating that the installation is in compliance, should satisfy the requirements of the insurance company.

Another issue for the do-it-yourselfer is service if the stove breaks down. Most pellet stove manufacturers, that deal directly with big box stores or online vendors, have a technical staff that can be reached by phone to help purchasers get through problems.

During the heating season, as service calls increase, you may have to wait to talk to a tech. There are times when you will need hands-on technical help. In these cases, a NFI Certified Specialist can help. You should keep the name and phone number of your local specialist handy.

It has been my experience that local pellet stove retailers will not service any stove they did not sell.

But what about buying a "used" pellet stove?

I would strongly advise against this course unless you purchase a unit from a local pellet stove store with some kind of warranty, and a commitment to offer service if there are problems with the stove.

Purchasing a used pellet stove from an individual leaves you without recourses if the stove needs service or repairs. I would rather see someone buy a base-line pellet stove from a big box store than buying a used one. When you buy new, you get the benefit of the technical support of the manufacturer. Keep in mind that parts for used, older stoves may no longer be available.

Chapter Six
Summary and Checklist

1. **Get the necessary approvals from your local and state officials.**

2. **After deciding on a particular stove, go online and review the owner's manual of that manufacturer for the model you have chosen.**

The manual will contain a summary of the codes and standards that you must comply with for the stove location and clearances to combustibles. It will also show the venting requirements for that particular stove. Remember there are special requirements for mobile homes, and most manufacturers do not permit pellet stove in bedrooms.

3. **Have some awareness of how a particular stove operates.**

Here again, the manual will give full instructions on how to use the control board to operate your stove.

A general idea of the components of a pellet stove will be helpful

THE FUEL DELIVERY COMPONENT

Pellets are delivered from the hopper to the burn pot via the auger.

THE COMBUSTION AIR DELIVERY COMPONENT

Air is drawn into the combustion chamber and to the burn pot by a combustion or draft fan that blows air into the exhaust vent. The action of the combustion fan creates a negative pressure in the combustion chamber that must be maintained for proper operation of the pellet stove.

THE HEAT TRANSFER COMPONENT

Heat is transferred by *conduction, convection* and *radiation.*

All three are evident in pellet stoves. A convection or room fan is employed to blow room air through and over the heated surfaces of the stove, delivering warmed air into the living space.

THE ASH COLLECTION COMPONENT

The combustion of pellets in a pellet stove produces ash that is collected in either ash boxes or drawers located below the combustion chamber or in ash pans located in recesses in the bottom of the combustion chamber. Ash so collected, must be periodically emptied.

THE VENTING COMPONENT

The National Fire Protection Association (NFPA) requires special venting, known as PL or L type venting for pellet stoves. No other venting is permitted unless approved by the Pellet Stove Manufacturer.

The owner's manual of the manufacturer will specify the total horizontal and vertical length of the venting and number of Ts and elbows permitted.

4) Become familiar with the periodic servicing and cleaning of your pellet stove as per the manufacturer's instruction's found in the owner's manual.

5) If possible, buy your pellet stove from a pellet-stove store where professionals will install and provide technical help when needed.

If you decide to purchase a new pellet stove from a big box store or online, plan ahead, taking into account the CODES, STANDARDS and manufacturer's instructions.

6) You must notify your homeowner's insurance company.

If you need help in planning or technical support, go to the website of the -
NATIONAL FIREPLACE INSTITUTE
(www.nficertified.org)

7) Buying a used pellet stove is not a good idea unless you purchase it from a pellet stove store and you can get some kind of warranty and also a commitment they will available for technical support.

Final thoughts…

I realize that I have given you much to consider; but for all the details and bother you will be glad that you have decided to purchase a pellet stove.
I hope you will enjoy it as much as I enjoy mine!

Thanks

Andrew

Purchased from-	
Date of purchase	
Make	
Model number	
Serial number	
Date of manufacture	

Tech support phone number for Pellet Stove NFI Certified Specialist

_____   _____

Name   Number

_____   _____

Name   Number

ANNUAL MAINTAINENCE CHECKLIST

Empty the pellet hopper	
Let stove run and burn out all the pellets in the auger tube. Allow the stove to shut down by itself.	
Unplug the stove for the season	
Vacuum the hopper of all pellet dust	
Make sure air intake is clean	
Vacuum the combustion chamber – including the heat exchanges. There are baffles that must be removed, ash traps that are exposed and must be vacuumed.	
The burn pot must be cleaned, air holes cleared, and any creosote build up scraped clean. (This is also an important part of daily and monthly maintenance)	
The exhaust channel or tube, running from the combustion chamber to the combustion or draft fan, must be cleaned of all ash build-up, and any creosote scraped clean.	
The combustion or draft fan housing must be opened and the motor fan blades scraped, removing any soot or creosote. (A combustion fan with crusted fan blades will slow the airflow and rob efficiency)	
Vacuum the fan housing and exhaust passage to the venting.	
Empty the cleanout Ts, and using a pellet-stove brush, sweep out all parts of the venting.	
Record the Date of your completed maintenance here	

Notes

Made in United States
North Haven, CT
29 September 2023

42146263R00036